Kitten Care

Photo Credits

Robert Pearcy: 1, 5, 6, 7, 9, 19, 23, 26, 27, 29, 39, 45, 49, 53, 56
John Tyson: 3, 10, 24, 50
Richard K. Blackmon: 4
Jacquie DeLillo: 16, 61
Gillian Lisle: 20
Isabelle Francais: 21, 32, 43, 48
Vincent Serbin: 31
Joan Balzarini: 46, 54
Ron Reagan: 59

T.F.H. Publications, Inc.
One TFH Plaza
Third and Union Avenues
Neptune City, NJ 07753

This book has been published with the intent to provide accurate and authoritative information in regard to the subject matter within. While every precaution has been taken in preparation of this book, the author and publisher expressly disclaim responsibility for any errors, omissions, or adverse effects arising from the use or application of the information contained herein. The techniques and suggestions are used at the reader's discretion and are not to be considered a substitute for veterinary care. If you suspect a medical problem, consult your veterinarian.

Library of Congress Cataloging-in-Publication Data
Quick & easy kitten care.
p. cm.
Includes index.
ISBN 0-7938-1029-9 (alk. paper)
1. Kittens. 2. Cats. I. Title: Quick and easy kitten care. II. T.F.H. Publications, Inc.
SF447.Q53 2005
636.8'07--dc22
2004024167

www.tfhpublications.com

Table of Contents

You and Your Kitten

Congratulations on your decision to bring home a kitten! You're in popular company. Felines replaced canines as America's number one pet several years ago, and their popularity and numbers have steadily soared. There are approximately 75 million cats living among us and with us in our households. Obviously, there is something agreeable about having a cat.

Taking on the responsibility of raising a kitten and caring for him for the rest of his life—which could possibly be for 20 years or more—is what you must consider before bringing a new kitten home. Raising a kitten is not a job to be taken lightly. Kittens are

Pet Population Continues to Grow

The results of the 2003-2004 American Pet Products Manufacturers Association survey show that the US pet cat populations continue to grow. The number of pet cats has increased from 72 million in 2000 to 77 million, while the number of dogs decreased from 68 million to 65 million. Obviously, cats are still "top dog"!

rambunctious. They are discovering the joys that the world has to offer, and everything is brand new to them. What an amazing revelation when your kitten finds out that the roll of toilet paper will go around and around and is perfect for shredding!

The perfect age to acquire a kitten is after he has been weaned from his mother and can eat solid food on his own, which is usually after eight weeks. Taking a kitten away from his mother and siblings too early can result in an emotionally handicapped animal who may have not been taught essential behaviors, such as grooming.

Once the kitten has been taken away from his mother and littermates, he may feel a little lonely and abandoned (after all, he is on his own now). This is one reason you need to step in and make the kitten feel like a loved and valued member of the household. Be sure to play with your kitten every day, several times a day. (If you adopted two kittens, this will be easy; they will constantly play together!) He

Even this young kitten displays the grace and beauty that has made the cat the most popular pet in America.

will welcome the attention, and you will be forming a bond with your new pet.

Choosing the Right Kitten for You

Your kitten may demand more interaction from you than an older, mellower cat would.

Knowing how to choose the right kitten for your lifestyle is essential. Besides yourself, you should consider other household members (including other pets), how active the household is, and other factors such as size (when full grown), coat length, purebred or mix, gender, and temperament.

Pet or Show Cat?

The first thing you should consider is whether you want to purchase a purebred, or pedigreed, kitten and compete in shows, or whether you really just crave the companionship of a good friend. While breeding is a legitimate business for most, some "backyard" breeders are only breeding felines to make money off them by selling the kittens or perhaps winning money at a cat show. Buying a kitten from this type of breeder will contribute to the vicious cycle of unwanted litters of kittens and cats who end up in shelters across the country. You must be very careful when buying a kitten from a breeder; get references from other customers who have bought kittens from him or her.

Even if you do decide to purchase a purebred (with papers), the kitten may not be show material because of behavior problems. Some felines like to strut their stuff in front of people, but others are too nervous to meet the demanding schedule and pressure of being a show cat.

Male or Female?

Another consideration is the sex of the kitten. This is a bit of a controversial topic; experts cannot seem to agree if sex actually makes a difference in the personality and behavior of an adult cat. The alteration (spaying or neutering) of a kitten makes an obvious difference in both sexes because of the mating instincts it curbs, but is there a difference between the altered female and male?

In general, many males seem to be more affectionate and loving, while females are often more aloof and standoffish. It is important to remember that actual scientific evidence in the field of behavior patterns exhibited by altered cats is very difficult to find. The bottom line is that once you get to know and love your kitten, personal preference always outweighs any statistics.

Purebred or Mixed Breed?

Purchasing a purebred kitten is not to be entered into lightly. Do you homework on all the different breeds available, and find a reputable breeder (any of the cat registry associations, such as the Cat Fanciers' Association, can direct you in your search) to work with and find the right breed for you. Working with a purebred is not for everyone, however, and you may just want to find a loving mixed-breed kitten in need of a good home.

If you decide to buy a purebred kitten, finding a reputable breeder is important because you want to know that your "investment" (and

How to Spot a Healthy Kitten

- Clear, bright eyes
- Pink mouth and gums
- Clean-smelling ears
- Silky, smooth coat
- Appropriate weight for age
- Active and alert to surroundings

Your kitten will be just as fun and lively a companion whether he's a purebred or mixed breed.

purebred kittens are usually expensive enough to be considered as such) is healthy and well socialized. This means the kitten should have lineage papers, vaccinations (if possible), and been handled by humans quite frequently for socialization.

Alley cat, Moggie, non-pedigreed, mutt-cat, and mixed breed are just some of the terms used to refer to felines who are not bred from any certain pedigree—who have no papers tracing back their lineage or whose parents (one or both) were not purebred. But these are the kittens and cats most of us know—the stray and feral felines hiding in the cornfield across the road, or perhaps the faces staring at you from behind the bars at the hundreds of animal shelters across the country.

In the end, you won't care where your kitten came from. The important thing is that he's with you, sharing the joys life has to offer.

One Kitten or More?

Cats are by nature solitary creatures—that's part of their fascination—and, with the exception of a few extremely sociable breeds, can certainly live very successfully as "only cats," with just their humans for companionship. Having a single kitten may pare down food, litter, and veterinary bills, but the owner of a single cat also has a greater responsibility because of his or her role as chief companion and entertainer. Two kittens, especially littermates, bond closely, keep each other entertained, and are good companions for the hours you're away during the day. Introducing two unrelated kittens is usually not problematic, although both should be veterinarian-checked and isolated for the first few days to preclude infectious diseases.

For a true cat lover, a single cat only begins to ease the addiction. However, some people may feel that their apartment is too small for more than one or that they can't handle the added expense and time, including litter-cleaning chores.

Bringing two or more kittens into your life is about as easy as bringing one, and only slightly more expensive. Many people find that having two play together and keep each other company during the day alleviates their own guilt about being away for hours at a stretch. If there's already a resident cat, it's generally much easier to introduce him to a kitten than to another adult cat. In a worst-case scenario, if something should happen to one of your cats, you will not be totally bereft. Cat lovers find it's sometimes wise to have "an heir and a spare."

Having two kittens provides mental and physical stimulation for both when you can't be home.

Where to Find Your Kitten

Once you've decided to get a kitten and may have even

A Quick and Easy Test

One test to see how the kitten moves is to stand a few feet away and call him to you. This will let you see how the kitten walks (you'll notice any limping, stiffness in the joints, or balance problems) and will also give you an idea of how interested the cat is in you.

made up your mind about what type you're looking for (such as breed and gender), the time has come to search for your perfect kitten. Keep in mind that your choices are many and varied, and that all kittens are adorable. Regardless of where or from whom you get your kitten, be sure it's healthy.

Selecting a Healthy Kitten

It may sound simplistic, but a healthy kitten looks healthy. Your

Kitten Free to Good Home

Searching through your local newspaper's classified advertisements is a common way to find kittens. In most cases, the felines are listed as "free to good home," and there won't be a price attached. Usually the person placing the ad in the newspaper wants to get rid of a pet because he or she is moving or was surprised by an unexpected litter of kittens.

You should thoroughly check out the kitten in question, however, and make a special note of the surroundings. Are they clean? Does the kitten seem to be in good health? Be sure to get a history of the kitten and ask questions. Will the person take the kitten back if a veterinary exam shows that something is wrong? What vaccinations has the kitten had? Be sure you feel comfortable with the kitten and owner before you bring the cat home.

Kittens, Kittens, and More Kittens

A fertile female cat can produce up to three litters in a year's time. The average number of kittens born to a mother cat is usually somewhere between four and six per litter. One female cat (and her offspring) can produce approximately 420,000 kittens in only a seven-year period.

potential pet should have a well-groomed, alert, clean appearance overall.

The kitten's coat should be clean, fluffy, silky smooth, and well maintained, with no missing patches. Feel the kitten's entire body for any lumps or hard spots. Take a moment and check for fleas. Any kitten can get fleas, and finding fleas is not a reason to avoid adopting a kitten, but keep in mind that the kitten will need to be given a veterinarian-approved flea treatment.

The kitten's eyes should be clean, clear, and free of runny discharge or tearing. Make sure the kitten's eyes focus on you and are able to follow an object moving in front of him.

The kitten's ears should be clean and free from waxy buildup. Dirty or smelly ears and a constant shaking or scratching of the head are often signs of ear mites (which are easily treated, but still unhealthy).

Check the inside of the kitten's mouth. It should be pink and smell clean (unless, of course, he just ate). A kitten's "baby" teeth (or milk teeth) should be white and bright. These teeth grow in around two weeks of age. The kitten will lose his milk teeth when he is between four and six months.

The kitten's nose leather should be dry to the touch, and there should be no discharge.

The area around the anus and the base of the tail should be clean and dry, with no signs of diarrhea.

Once you know what to look for in a healthy kitten, it's time to decide where to find him.

Strays

There are two types of "stray" felines. A friendly stray is one who had a home previously, while a feral stray is a wild feline who neither likes nor trusts humans.

Finding an abandoned stray sometimes almost seems like fate. After careful examination by your veterinarian, a homeless stray can become a very rewarding part of your life. Your new kitten may be stressed out for a while, worried that you will leave him, but an abundance of love and attention will soon put his fears to rest.

Feral cats and their kittens have had hard lives. You can provide an outdoor shelter and food and water for your new feline companion, but you will probably never enjoy the benefits of stroking the kitten on your lap. You also have to make sure that you trap the kitten and

Strays Among Us

Often you don't just find one lonely stray. It is very possible that you will find a stray mother with an entire litter of kittens. Irresponsible cat owners who do not spay their female felines may find it "easier" to dump off the pregnant mother cat somewhere instead of dealing with her and her litter. It is estimated that there are 60 to 100 million homeless felines in this country alone. Acquiring a kitten by chance, instead of by choice, is something that happens more frequently than not.

Tugging at Your Heartstrings

When you're looking for a healthy kitten, you may come across one that is not quite right for some reason. The kitten may be ill, suffering from a medical or physical condition, or may be an aggressive or nervous type. Unfortunately, a kitten like this is not a good candidate for adoption, especially for a first-time kitten owner. Although you may be tempted to take in and reform or rescue the kitten because you feel sorry for him, avoid the temptation. Unless you have the time and patience (and in some cases, finances) to devote to the kitten, you may be getting in over your head. A kitten with a chronic medical condition may need medication daily throughout his lifetime. Bringing a sick animal into your home could make other pets ill. New adoptees with personality disorders can upset the balance of the household if there are other pets present. If you adopt this problem kitten and later find out that you cannot care for him properly, what will you do?

However, if you feel you have the time and energy to devote to a kitten with special needs, then by all means take him home and give him all the love and care you can. But for the overall good of everyone involved, think the situation over carefully before making a final decision.

take it to a veterinarian for a complete checkup and to have the kitten spayed or neutered.

"Spare" Kittens

Many cat owners continue to let their females have litter after litter of kittens, then make the rounds to all their friends and family members to see who wants these "spare" kittens. Normally, getting a kitten from a friend or acquaintance is a safe bet because you know what type of environment the kitten had lived in and the health and personality of the mother cat. You will also have plenty of time to play with, hold, and pet each kitten so you can choose which kitten (or kittens) are best for you.

A Pet Store Kitten

If you are considering purchasing a kitten from a pet shop, it is important to observe the kittens. You can certainly tell a healthy, active, alert kitten from an unhealthy, sluggish, unfriendly one. Ask the pet shop personnel about the kitten's background, as you would any source from which you obtained your new family member.

Purebred Kitten from a Breeder

There are hundreds of breeders to choose from. Before buying your kitten, be sure to visit the cattery where the kitten was born. If the breeder is uncomfortable with you visiting, be cautious—he or she could be trying to hide something.

The cattery should look orderly and smell clean. The animals should be in clean, roomy cages or in a confined area. All of the cats should appear friendly and healthy. Signs of illness in any member of the group could be trouble, since cats commonly pass diseases and illnesses to one another.

Top Ten Most Popular Breeds

According to the Cat Fanciers' Association's (CFA) 2003 registration statistics, the top ten most popular breeds were:

1. Persian
2. Maine Coon
3. Exotic
4. Siamese
5. Abyssinian
6. Birman
7. Oriental
8. American Shorthair
9. Tonkinese
10. Burmese

Before you bring a kitten home, make sure that no one in the household is allergic and that everyone agrees to the decision.

Find a breeder who spends time with the kittens and handles them daily. Early socialization with humans is a crucial factor in a cat's overall personality.

Many purebred kittens come with a hefty price tag, so you will surely want to avoid getting an unhealthy kitten or one who does not meet the required standards if you intend to show him. Make sure you receive the lineage papers and certificates as well as proof of inoculations and other important medical information.

A Shelter or Rescue Kitten

An animal shelter is another place to look for a kitten to adopt. All shapes, sizes, breeds, and temperaments of kittens are available through animal shelters, rescue organizations, and humane societies. If possible, visit several before making your choice, so you are sure of choosing the best kitten for you and your family.

Your Kitten at Home

The first day your kitten arrives in your home is an exciting one, but before that day comes, you need to have everything ready for your newest family member.

Proper Supplies

Before your kitten's arrival, make sure that you have the proper supplies. You should be able to find these at a pet store near you, so take a shopping trip a few days before you pick up your kitten to minimize stress once he's in your home.

Be Prepared

Before you bring your new addition home, be sure to have the following on hand:

- Cat carrier or crate (to transport your cat home in)
- Litter box and scoop
- Cat litter
- Food and water bowls
- Kitten food
- Brush and comb (for grooming)
- Collar and ID tag
- Toys
- Scratching post
- Cat bed

Cat Carrier or Crate

You will need to bring your kitten home in something safe and sturdy. A plastic cat carrier or crate with a locking mechanism on the front will be secure enough. (A cardboard cat carrier or box is not sturdy enough, and a frightened kitten could easily escape or claw his way out.) The cat carrier should have adequate ventilation and allow you easy access to your kitten. Be sure to get a carrier that will be large enough for your kitten when he is full grown. The Nylabone Fold-Away Pet Carrier® is a good choice, as it is sturdy and durable but folds away for easy storage and convenience. Put a towel or old sweatshirt on the bottom of the carrier so your kitten has something soft to sleep on on the way home.

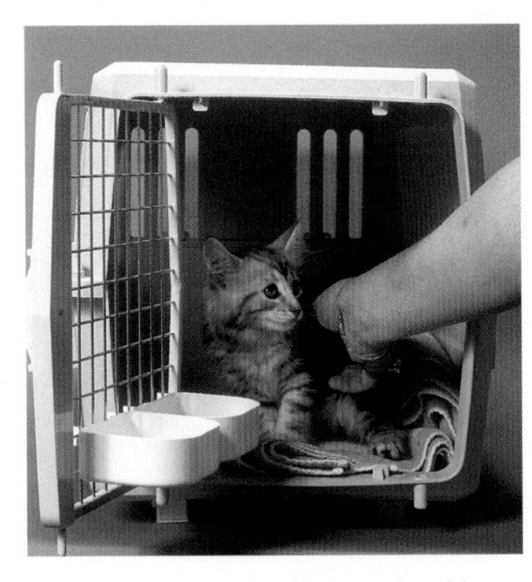

Provide your kitten with a comfortable carrier in which he can feel safe and secure.

Food and Water Bowls

There are several kinds of food and water bowls available. Steel bowls do not break easily and cannot be scratched, but some cat owners question whether the food and water pick up a metallic taste. Another option is heavy earthenware or ceramic bowls. Make sure the bowls are sturdy and have weighted bottoms so they cannot be tipped over. (Use caution if you buy a painted ceramic bowl; some paint may contain lead, which is harmful to your kitten.) Plastic food dishes are not a good idea for pets. Over time, the plastic can become scratched, and germs and other bacteria can live inside the bowl. To prevent fighting over food, it is a good idea to have one food bowl per cat in the household.

Clean the food and water bowls daily, and place them far from the litter box. The kitchen is a good place for a feeding station.

Food

Be sure to have an ample supply of kitten food on hand before your kitten comes home. Find out from the breeder or animal shelter what brand he's been eating, and stick with it. Ask about food preferences and whether he has any special dietary needs.

Provide your kitten with a perch or ledge from which to observe the outside world.

Bed

Although felines have a tendency to sleep wherever they are when they become tired, a cat bed will be a favorite spot to catnap.

Kittens feel safer if they can sleep in an elevated bed. You can buy any of several commercially manufactured cat beds in pet stores. Make sure the bed is large enough for your kitten to fit in comfortably but small enough to give him a sense of security. Place the bed in a semi-dark, quiet, out of the way area of your home so that your kitten can get all the rest he needs.

Litter Box

A litter box is a necessity, and there are many different styles to choose from. Make sure that you get a box small enough so that your kitten can get in and out with no trouble. Otherwise, he may decide it's easier to eliminate elsewhere. For very young kittens, the litter box should be very close to their sleeping quarters, but not too close to their food and water.

Scratching Post

All cats need to scratch—it's part of their nature. Make sure that the scratching post you buy has a sturdy base so it won't tip over. Get one that is at least as tall as the kitten will be when he's full grown.

Kittens may chew on plants that are toxic or even poisonous to them, so make sure you've removed any potential dangers before your kitten arrives.

A curious pet can suffocate in a plastic bag, so be sure to keep these potential hazards away from your kitten.

That way, he can reach up to his full height and get the best scratch possible. If you have a big house or more than one cat, get several scratching posts so the felines can scratch whenever they get the urge. Keep the scratching post attractive to your kitten by "refreshing" it with catnip scent.

Toys

There are thousands of types of cat toys on the market, and each kitten will have his favorites. Your primary concern is to make sure that any toy you give your kitten is safe and contains no small parts that can be torn off and become choking hazards. Items such as bells, feathers, and pom-poms can easily be ripped off while your kitten is playing with them. Never give your kitten a toy that is small enough to be swallowed.

Collar, ID Tags, and Leash (or Harness)

All pets should have a collar and identification tag with your name, address, and telephone number on it. Make sure you get a collar with a release that will prevent your kitten from getting hung up on

Kitten Safety

A kitten needs to be taught where he can and cannot go, for his own safety as well as your peace of mind. You don't want your kitten getting into places he can't get out of (such as a dresser drawer or behind the refrigerator), nor do you want him climbing up the back of your sofa. Kittens are much like human toddlers; they can and will get into everything possible.

The best way to kitten proof your home is to think like the kitten. Get down on his level and crawl around on the floor to see what looks tempting. (It sounds a bit silly, but you'll notice the same things your kitten will.) Kittens will play with anything, and all too often, they will try to eat their toys. All potential choking hazards need to be removed from the clutches of their tiny paws. Be sure you keep a close eye on your little bundle of fluff and watch out that he doesn't get into trouble exploring. Correct him with a gentle "No" if he goes somewhere or does something he shouldn't and redirect his attention on something positive.

a tree branch if he goes outside. If you are going to train your kitten to walk on a leash, buy a secure, escape-proof harness to go with it.

Kitten Proofing Your Home

It is best if you remember to treat your new kitten as you would a human toddler entering your home for the first time.

Before your kitten comes home, put away anything that could prove dangerous to him. Kittens are extremely curious and will eventually examine every nook and cranny of your home. Be sure that anything that is toxic is put away in a safe location. Don't forget that many kittens learn how to open cupboards and pantry doors. Nothing can be considered safe if you think of your kitten as a toddler—medicines, cleaning supplies, sharp objects, etc. must all be out of reach.

Everyday household items can be dangerous, too. Simple things such as lit candles, cleaning products, plastic grocery bags, rubber bands, and pen tops can all cause him harm.

The First Few Days at Home

Remember that your kitten will need time to adjust to his new surroundings. Your house, family members, and other pets all seem strange to him. Be patient with your kitten for the first few weeks while he becomes used to living in his new home.

Everyone is excited when a new kitten is brought home. Friends and family members may want to come over to see him and play with him. Your children may want to hold and cuddle with him. At this time, however, you need to provide your kitten with just the opposite.

Your kitten will most likely be afraid and stressed when you bring him home. He will miss his littermates and may cry for his mother, so you must be patient and loving and soothe away his fears. He

Make sure to kitten proof your home before your kitten arrives, or else your new charge may get into mischief.

Popular Plants That Are Toxic to Cats

Although there are many plants that can prove toxic to your kitten, this short list is just a few of the more well-known—or popular—plants that are often found in the homes of plant enthusiasts.

Aloe Vera, Amaryllis, Azalea, Baby's Breath, Bittersweet, Bleeding Heart, Bluebonnet, Chrysanthemum, Daffodil, Easter Lily*, Eggplant, English Ivy, Eucalyptus, Ferns, Geranium, Heartland Philodendron, Holly, Honeysuckle, Iris, Marigold, Mistletoe, Morning Glory, Oriental Lily*, Peony, Periwinkle, Poinsettia, Primrose, Rhododendron, Rubber Plant, Schefflera, Tiger Lily*, Tulip, Wisteria

*Lilies are especially dangerous to cats and should be avoided at all costs!

may be startled by anything loud and different (like a vacuum cleaner or the television). Do all you can to make sure your kitten feels as comfortable and secure as possible until he has settled in.

Prepare a special room for your kitten so that he has time to adjust to his new, unfamiliar surroundings.

Prepare one room (such as a spare bedroom or bathroom) where your kitten can spend his first day or two. Provide him with a soft, warm cat bed, toys, a litter box, food, and water. Let him live in this special room until he has adjusted to being on his own. A quiet, safe place will help your kitten into his new home faster. Talk to your kitten calmly as you pet and play with him.

After he has adjusted to his special room, you can let your kitten explore other parts of the house. It is a good idea to start out slowly. Let your kitten explore one or two rooms a day until he feels comfortable, then allow him into other parts of the house as he relaxes in his new environment.

Introducing Your Kitten to Your Other Cats

A proper introduction is crucial when familiarizing your kitten with any other cats in your home. Felines are territorial, and very envious when the limelight shines on anything other than them. It is important that you constantly reassure your other cat (or cats) that they are not being replaced but instead have the chance to make new friends.

When introducing your kitten to your first cat, you need to always be present in case fighting occurs. This is especially true if your first cat is an adult. When you cannot be there, one good way to get the felines used to each other is to keep them in separate but adjoining rooms so they can at least smell each other and touch paws underneath a closed door. No one can get hurt this way, and the felines should learn to accept each other.

You can also use a travel crate to introduce your new kitten to your first cat. Set the crate with the new kitten in it on the floor. Your

The First Few Days

Spend extra time with your new kitten for the first few days until he has started adjusting to his new surroundings. If possible, bring him home before a weekend or arrange to have time off from work to stay with him. You don't want to leave your new friend alone, afraid and lonely, in a strange place for hours on end. The more time you spend with your kitten, the stronger the bond of love will be between the two of you.

Older cats may more readily accept a kitten than another adult; simply make sure to introduce them in a safe environment.

established cat will most likely come over to investigate the new addition to the household. After a short time, switch the felines; put the established cat in the crate and let the kitten wander around the room. This method will allow the two to see and sniff each other and will prevent fighting.

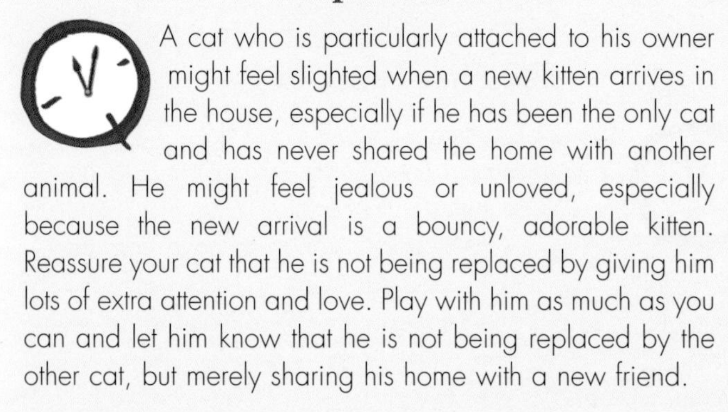

Replaced?

A cat who is particularly attached to his owner might feel slighted when a new kitten arrives in the house, especially if he has been the only cat and has never shared the home with another animal. He might feel jealous or unloved, especially because the new arrival is a bouncy, adorable kitten. Reassure your cat that he is not being replaced by giving him lots of extra attention and love. Play with him as much as you can and let him know that he is not being replaced by the other cat, but merely sharing his home with a new friend.

Holding

Teach your family members the proper way to hold a kitten. Hold the kitten under the chest and support his feet with your arm or hand so his legs do not hang free. This will give your kitten a feeling of comfort and stability.

The world of felines has a hierarchy. There is always a cat in charge who is known as the "alpha." Usually the dominant cat is a male, although if all the cats are altered, a female may hold the dominant position. When felines are introduced, one will usually know immediately whether he is the dominant cat. The submissive feline, when meeting the "top cat," will stand his ground for a minute or two worth of howls and growls, but he will eventually slink off to sulk.

One of the best ways to get two or more felines to like (or at least tolerate) each other is to play with them together. Get out the cat toys and play until your established cat forgets that his space has been invaded. Soon, your two felines will be playing with each other on their own.

If you have multiple felines in your home, it is a good idea for each one to have his own litter box, food dishes, and toys. Even if cats get along fine with each other, they may not want to share their toys or stand in line at the food dish.

Kittens generally get along well with kids, and vice versa, but be sure to train children—your own as well as visitors—to handle your kitten properly.

Some felines will not use a litter box that another cat has used because it has the scent of the other cat on it.

Kittens and Children

Kittens and children can safely coexist in the same household as long as certain rules are followed. Small children need to be taught what not to do (such as pull the kitten's tail), how to hold a kitten properly, and to respect the kitten as an individual being with feelings.

Children should be supervised whenever they are with the kitten. In addition to showing the child how to pet the kitten, also teach the child that sometimes the kitten would like to be left alone. Since both kittens and young children are somewhat unpredictable, leaving them alone together is not a good idea. If you are there to teach your child that the kitten is not a stuffed animal that can be handled roughly and treated carelessly, your child and your kitten will get along fine.

Feeding Your Kitten

Feeding your feline the correct diet is something that you should think about at every stage of your cat's life. Every cat demands different nutritional requirements, and you need to know what these dietary needs are and when they should change.

Only you can ensure that your kitten grows up to be the strong, healthy cat he is capable of becoming—both physically and mentally. His first year is an important one. With proper diet and care, you can help start your kitten off on the road to good health. The correct diet will help your kitten develop strong bones, good teeth, and a lustrous coat.

Can Cats be Vegetarians?

Cats cannot survive on a vegetarian diet. In 1982, taurine, an amino acid only present in meat, was found to be crucial for the healthy functioning of the cat. Taurine is vital for aspects of metabolism, eyesight, cardiac function, bile formation, and reproduction. Cats are not able to produce sufficient taurine themselves and must have an adequate dietary supply. Since taurine occurs almost solely in materials of animal origin, cats should not be fed on a purely vegetarian diet.

When your kitten does start to eat solid foods, consult with your veterinarian about the best type of food for his particular needs. Most pet food manufacturers make kitten foods in dry, semi-moist, and canned formulations. It is important that you feed your kitten a food that offers a 100 percent complete and balanced diet. If the food your veterinarian approves is different than that which the kitten was fed at the breeder's or the animal shelter, ask your veterinarian the best way to introduce the new food into his diet. Usually, it's best to mix his old food with his new food. Gradually

Cats cannot survive without meat, since they are unable to produce taurine, an essential amino acid.

increase the amount of new food added to the old food. Over time, your kitten will be eating only the new food and will not have noticed the gradual change. Hopefully your kitten will not miss his old food, but if he does, you may have to slow down the conversion process. Mixing the food until your kitten adjusts to a new diet will make the conversion much easier and less stressful for you and your kitten.

Give your kitten small, "mouse-sized" meals three to five times a day.

Schedule

Kittens are much like human babies. They need to eat and drink often—three, four, or five times a day. Serve your kitten little portions; don't let his stomach start to growl from a lack of food, but don't serve him king-size portions either. Ideally, cats like small "mouse-sized" meals. For a kitten, you might want to put out three to five small meals a day, with premium dry food available for snacking. This may vary depending on which formulations you decide are best for your kitten, and the frequency of meals will lessen as your kitten grows.

Follow the instructions on the kitten food bag or can to get a general idea of the correct amount of food to give your kitten. You can then adjust this amount to meet your kitten's exact requirements. A lean, well-conditioned body is the ideal.

Each kitten or cat should have his own bowls: one for food and one for water. Wash your kitten's bowls in hot soapy water after each feeding, just like you would do your own. It is easy to forget to do

The Milk Myth

Many people believe that a bowl of warm milk is just what a kitten needs to lap up before taking a long nap. However, cow's milk is the last thing your kitten should have for a meal, since most felines are lactose-intolerant. Although they may greedily lap up any milk that you offer them, they cannot digest it properly, which may lead to a bout of diarrhea. Diarrhea in a small kitten can quickly lead to dehydration and in the worst-case scenario, death. It is best to avoid giving your kitten any milk, no matter how much he may seem to love it.

this, to get lazy, and instead just simply add food to the food left behind from the previous meal, but doing so can lead to upset stomachs and finicky eaters. Do not try to save leftovers until the next meal, even if it means that some of the food is wasted. Of course, dry food can be left in the dish a little longer, but it too should be discarded each day.

Provide your kitten with water, not milk; dairy products can actually upset a cat's digestive system.

Set up a routine, just as you'll establish one for cleaning the litter box. At the end of each day, make sure that the bowls are clean and that the drinking water has been replenished. When you get up in the morning, serve the first meal of the day in a clean bowl.

Feeding Guidelines

It may be tempting to feed your new kitten with foods that you like, but do not be fooled by your kitten's apparent delight in table scraps and other goodies. Your kitten's diet requires nutrition specifically designed to meet his needs; if he is fed other types of food, these very important needs will not be met. In addition, if you give him a chunk of food one day, he'll expect it day after day. Felines do not rationalize like humans do. Your kitten will fully expect that if he can eat your food once, he can eat it all the time.

Cats and Grass

Most cat owners have witnessed, at one time or another, an unwell feline snacking on outdoor grass. This seems to have a medicinal purpose for them. Experts know that grass can cause a cat to vomit, which helps bring up excess fur that's been swallowed.

Some cat owners will pick grass from their lawn (make sure that no lawn chemicals have been added to kill weeds) and bring it inside for kitty to graze upon at his leisure.

You can also buy a special kit of "cat grass" at your local pet shop or in most grocery stores. The kit comes with seeds of grass that are safe for your cat to eat. This is one way of bringing the outdoors indoors for your feline friend. Whether planted and grown in a kitty container or plucked from a chemical-free lawn, providing greenery to supplement his diet is a generous gesture your cat will surely appreciate.

There are two options when considering feeding your kitten. Some people prefer to leave food out 24 hours a day so that their kitten can nibble any time he feels hungry (this is called "free feeding"). You should note that such feeding should only be done with dry food to minimize the risk of spoilage. Other people, however, fear that their kittens will become overweight if they are allowed free access to food all the time, so they feed their kitten on a schedule.

Some kittens and cats will continually overeat if they are allowed free access to food throughout the day. Also, if you have more than one feline, you may find free feeding difficult because of the competition between them. In this situation, adopt a portion-control feeding routine. Simply pre-measure the quantity of food that your kitten should receive for the day, then feed half in the morning and half in the evening. (It should be noted that these feeding options are for kittens who are at least eight weeks old.)

Water

Your kitten should have access to clean fresh water at all times. Even if it appears that your kitten is drinking no water, it still needs to be available, because you may not see the kitten take a few drinks every now and then.

Water is an essential part of any animal's diet. If you are not sure about the taste or chemical composition of your tap water, either

Taste-Testing Felines

Did you know that when your kitten takes a drink of something, he usually doesn't swallow any of the liquid until after four or five practice laps? It seems that kitty is testing things such as the temperature and quality of the beverage he is about to drink before taking the actual plunge.

use bottled natural water (which can be found at any grocery or convenience store) or install a filter for your tap water. In addition, the water bowl should be thoroughly cleaned on a regular basis to prevent any buildup of bacteria.

If your kitten won't eat, it may be because you've switched foods or he's had an emotional upset—or he could simply be spoiled.

Finicky Felines

If your kitten will not eat, the first thing you should do is find out if he is ill. Sick kittens do not have much of an appetite, and your kitten's finicky palate could be the first clue that something is not right with him. A trip to the veterinarian should be your first stop in correcting a finicky disposition.

If everything checks out okay at the veterinarian's office, consider some common reasons why felines are (or turn) finicky. Believe it or not, your kitten is not going to starve himself (no matter what kind of impression he might be trying to give you).

One of the most common reasons for felines to become finicky and refuse to eat their food is that their food brands or flavors have been switched. Unlike humans, who enjoy having a variety of food to choose from, felines are very routine-oriented and dislike change of any kind. Unless your veterinarian recommends changing your kitten's food from the one that was originally approved, you really shouldn't. He will have to move to an adult formulation soon

Do Felines Have a Sweet Tooth?

If you've ever offered your dog a cookie, you know it will be gobbled up in an instant. The same is not true, however, with your cat. The reason? A cat does not have taste buds that detect sweetness; therefore, he can't be tempted with sweet rewards. Those finicky felines can, nevertheless, discern the slightest peculiarities in water.

enough. If your veterinarian does recommend making a switch, discuss the correct way to do it. Usually it is by mixing the new with the old, as discussed earlier in this chapter.

Just like a human, sadness and emotional upset can also cause an animal to lose his appetite. If something in your household has changed, such as a new person moving in, or even a person or other animal moving out, your kitten may appear finicky for a while because of emotional upset. Try to help him through this rough time and offer him something you know he will eat (such as a favorite treat) to get him started again. If your kitten doesn't start eating after a day or two, consult your veterinarian, or an animal behaviorist, to try to compensate for the changes (if possible) that caused the upset in the first place.

Another thing to consider is that your kitten is not really all that finicky, but rather is just spoiled rotten (not something that most pet owners want to admit). He may be so good at getting his own way that with one sad look at his food bowl, he can have you preparing an entire fish dinner. The solution is simple: do not let him have his way so often, and he will eat the food he is supposed to eat.

Whatever the reason for your kitten's finicky behavior, as long as you find the underlying cause, you can most likely deal with it and get your kitten eating once again.

Training Your Kitten

Litter Box Training

Although you may think you have to train your kitten to use the litter box, the one who really needs to be trained is you. A kitten has a propensity for cleanliness, which he learned from his mother, and an instinct to bury his waste in soil or litter. Once kittens leave their mothers, problems usually occur because the kitten's owner fails to keep the litter box and the litter inside it clean or fails to clean up the area around the litter pan. The number one rule in litter training is cleanliness. If you let the litter box get dirty, your kitten will let you know it—either by using the floor next to the box or by finding another, more sanitary spot in the house.

Make sure the sides of the litter box you provide your kitten with are low enough for him to step over.

Whatever the size or type of litter box, you must be able to clean and deodorize it efficiently. Scoop it out at least once a day, and wash it out on a regular basis, depending on how vigorously your kitten uses it (this could be anywhere from once a week to once a month). Cleaning the box with soap and water will do just fine, but be sure to rinse thoroughly so there is no soapy residue. Never use chemical cleaners, ammonia, pine oil, etc. Your kitten uses his paws in the litter box and will later clean them with his tongue.

Location

When you introduce your new kitten to your house, the first place you should show him is where the litter box is located. Once you've done that, do not move it. Pick a spot and leave it there. But where should that spot be? A busy, noisy area may make a shy or modest kitten decide to hide (such as behind the couch or under a bed) to do his business. Putting the litter box in an out of the way place to begin with will solve many problems before they have a chance to start.

Bathrooms are common locations for litter boxes, as are closets, spare rooms, basements, or any place that is not in direct contact with your family's daily activities.

Depending on the size of your home, you may need more than one litter box. If your kitten is at one end of the house when the urge strikes and the litter box is up two flights of stairs at the other end of the house, he may have an accident before he gets to the box.

For multi-feline homes, the general rule is one litter box per cat; most felines do not like sharing their box.

Litter

The type of litter you use is as crucial as the box itself. With very young kittens, it is probably better to stay away from the clumping types of litter for a while. There have been instances where tiny kittens could not properly groom themselves and have had clumps stick to the fur on their behind (which sometimes can cause an obstruction and be painful).

If your kitten won't use his litter box, try moving it to a different location.

Which Litter is Best?

If you're unsure which type of litter your kitten would prefer, why not set up several boxes, side by side, with different litter types in each? This test will allow the kitten to make his preference clear, so you can oblige him in the future. Litter box training can be simple and stress-free for both you and your kitten if you follow his lead. If you have trained your kitten correctly and followed all of the litter box rules and regulations, there should be a happy kitten in your household.

It is important to remember that cats are creatures of habit. There are many different brands and types of kitty litter. What you choose is up to you, but be aware that the first litter you use may condition your kitten to associate only that kind, style, smell, texture, etc. with the kind of litter suitable for eliminating. If you find a good brand of litter that your kitten likes (i.e., uses), your best bet is to stick with that particular litter to avoid problems later.

Accidents

There will be accidents at first, especially when your kitten is finding his way around his new home. He may accidentally soil carpets as he runs around. Make sure you clean and deodorize these spots; otherwise, they will be re-used for the same purpose. Use an odor-neutralizing cleanser to be sure your kitten or other cats don't resoil the area.

It is absolutely reprehensible to rub your kitten's nose or face in his accident to try to stop this behavior. Felines do not understand punishments such as this, and scolding your kitten will only encourage him to hide his accidents to avoid further punishment.

Behavior

Kittens do not misbehave on purpose; they are merely acting on their natural instincts. In the wild, seemingly "bad" behaviors (such as scratching) are part of a kitten's daily routine. Your kitten needs to learn how to adapt his inherent wild desires, like eating your plants and scratching your sofa, to the more domesticated setting of your home.

Methods of Correcting Your Kitten

Correcting any feline should never involve physical punishment. Cats do not understand being slapped or hit and will only respond negatively to this type of punishment. What kittens do understand is a very loud, alarming noise (a hand clap or a firm "No!") when they do something "bad." The hard part is, of course, catching them in the act.

All kitten owners should understand that if they come home to find that the kitten has done something he shouldn't have (such as knocked over the garbage can and rummaged through it), there is nothing they can do to correct him. He'll have no idea why you're making a loud noise. You may also want to keep in mind that you were the one who left the kitten in a situation where such a thing could happen.

When using the loud-noise method, it is crucial to make your kitten understand what he has done wrong and the reason for the response. This requires time and patience on your part. For it to be effective, not only do you have to catch the kitten in the act, you also have to try to make sure that he doesn't see you. He needs to

Making Eye Contact

A kitten's eyes can reveal what he's feeling or thinking. The pupils of a feline's eyes can dilate up to five times their normal size when he is frightened or feeling threatened. In normal light, a happy cat should have small pupils that look like little slits in the center of his eye. Eyes that are half closed signify a contented, relaxed, and (usually) sleepy feline.

think that the noise is directly related to his behavior, not something that comes from you. Felines are far from dumb. If your kitten realizes the noise is from you, he'll soon figure out that the "naughty" behavior is okay when you're not around to witness it.

The Urge to Scratch

The main reason felines scratch is to shed the sheaths of the claws they have so that they can grow new claws. Don't be surprised if you find little pieces of your kitten's nails scattered around the scratching post; it's just the sheath that has been shed, which is normal. Felines also have scent glands in their paw pads, and scratching any object will mark it as part of their private and personal property.

Scratching is also a form of exercise for kittens. A scratching post or kitty condo designed to be scratched and clawed will provide him with the necessary equipment for such exercise. Be sure to keep a scratching post near your kitten's favorite sleeping spot, because most felines love to stretch out and sharpen their claws when they awaken. Also, make sure the post is tall enough for him to comfortably stretch his entire body (remember to take into account that he is still growing).

It is important to teach your kitten good scratching habits (that is, to use his post instead of your furniture or carpet). The more scratching posts you have, the easier it will be for your kitten to fulfill his scratching desires.

Putting catnip on the scratching post will help attract your kitten to it. To encourage him to play at his scratching post, you may want to dangle a toy around the post or manually put your kitten's claws on it and show him the scratching motion.

A final note needs to be added about scratching and declawing. One of the biggest complaints by feline owners is scratching. Although declawing is a permanent solution to the scratching problem, it is an inhumane procedure. Some veterinarians refuse to do the surgery necessary to declaw a feline. Besides making the animal utterly defenseless, it can also cause serious psychological problems. Many declawed felines fight back by biting; they also may become depressed or overly aggressive. Clipping your kitten's nails often and teaching him to use a scratching post are much more humane methods of deterring him from bad scratching habits.

If your kitten scratches the furniture, redirect him to a scratching post.

The Best Solution for Bad Behavior

Kittens can learn and understand things just as we can. Perhaps they don't use the same process as human beings, but if you are fair and consistent, they will eventually learn what they should and should not do.

There are many ways to redirect improper behavior. Conditioning your kitten into appropriate behavior should go hand in hand with making certain he does not have access to the "bad" thing when you're not around. For example, if you work away from home, you might want to confine your kitten in a room where he does not have access to the object of his inappropriate behavior. When you're home, keep an eye on him, and should he try to do the "bad" thing, use your correction (usually a loud noice to startle him, and never a physical punishment) and then guide him to the appropriate behavior, which should result in praise and rewards. Through consistency, praise, and guidance, the correct behaviors will replace the wrong behaviors. For a more in-depth discussion on the subject, *Cat*

A scratching post is a safe and low-cost alternative to the drastic measure of declawing.

Training in 10 Minutes (TFH Publications, 2003) is an excellent book for learning about how positive, reward-based training can have your cat doing all sorts of things for you—with both of you loving it.

Keep in mind that your kitten isn't trying to do things wrong on purpose; he may just be curious about the interesting "human" things in the house (especially since he is a young kitten learning all about the world). Your kitten may be bored and craving human interaction, especially if he is an only cat. If you do not provide ample entertainment for your kitten, he may create his own fun and games, at the expense of your favorite chair.

Be patient when correcting and training your kitten; he will need time to learn what is and isn't accepted in your home. By examining the reasons why your kittens misbehaves, you will be able to come up with simple solutions to your cat's behavior problems. Almost all "bad" behaviors can be corrected. Take the time to teach your kitten right from wrong.

Grooming Your Kitten

Felines exhibit hair growth and loss cycles similar to humans. Shedding is part of the normal hair growth cycle in cats, and is directly affected by photoperiod changes, or the amount of daylight in the area. Thus, indoor felines who live where lights are kept on even when it's dark outside shed differently than outdoor felines in seasonal climates. Usually, cats stop shedding in winter because there's less light. Indoor cats never really stop shedding; they just do it all the time in lesser amounts.

Cats usually keep shedding mostly under control on their own, since they are naturally clean and constantly bathe themselves. If

Too Much Shedding?

Although shedding is considered a normal part of feline living, excessive or abnormal shedding could be caused by an illness. Ringworm is one such disorder. If you think your kitten is shedding an abnormal amount of fur, have him checked by your veterinarian.

you have a cat-hair problem, however, the most important thing you can do to decrease the amount of fur floating around your home is to institute a regular grooming schedule. Whether you comb and brush your kitten at home or use a groomer for the task, grooming can help a lot. Your long-haired kitten will be much happier if he has help removing the excess hair that can cause serious problems, including hairballs that can lead to vomiting and/or constipation.

Grooming Your Kitten at Home

Cats spend 30 to 50 percent of their waking hours grooming themselves. You may be wondering, then, why they need any assistance at all. While most short-haired cats do just fine with only an occasional grooming—if that—to keep household hair levels under control, some long-haired breeds cannot cope with all that fur on their own, no matter how diligent they are. Depending on how thick your kitten's coat is and how well he is able to manage it on his own, you may need to assist him anywhere from every week to every day.

Cats are generally content to groom themselves, but most enjoy being groomed by their owners as well.

The key to helping a feline accept your assistance with

The Long (and Short) of Cat Hair

Even though short-haired breeds require much less grooming, the largest number of cats registered in the US happen to be long-haired! According to the Cat Fanciers' Association (CFA), the world's largest purebred registry, only 10 out of the 40 eligible breeds are classified as long-haired cats. Although it would seem that the short-haired feline is more popular, nothing could be further from the truth.

Long-haired cats may have hairs up to five inches long, while in contrast, a short-haired feline may have hairs measuring less than two inches. CFA statistics from 2003 counted over 20,000 very hairy Persians registered in their association, accounting for nearly two-thirds of the total number of CFA-registered cats.

CFA's number two spot also goes to a long-hair—the mammoth-sized Maine Coon. What's surprising about the Maine Coon's numbers is that the statistics show only 4,300 registered—a startling difference of almost 16,000 felines between the number one and two popularity positions. About two-thirds of the total registered purebreds in the CFA are graced with a surplus of hair. Apparently, the long-haired varieties of our little furry friends are the decided favorites among feline fanciers.

grooming process is to start grooming him at a very young age. Starting a grooming regimen now, while he is still a kitten, will make it easier for both of you in the years to come. The best thing you may do for a kitten is to take out the comb and get him used to it, even before he really needs it.

Trimming the Nails

Before the grooming process begins, you may want to trim your kitten's nails to prevent getting scratched. A feline's claws should be

Grooming Supplies

For grooming a kitten, you will need:

- a brush or comb (the types differ for long-haired and short-haired cats)
- special nail clippers for cats (which you can get at a pet store or from your veterinarian)
- a little kitty toothbrush and toothpaste (which can also be found at a pet store or obtained from your vet)
- cotton balls
- styptic pencil

trimmed once a month. Place the kitten in your lap and, while securely holding her, gently press down on the paw so the nail comes out. With a pair of clippers designed especially for felines, clip off the tip of each claw just outside the pinkish part where the nerve (quick) is. Once the nails have been clipped, you're ready to groom your kitten.

If you accustom your kitten to grooming now, it will be a much more pleasant experience as he gets older.

Always have a styptic pencil on hand when trimming your kitten's claws. The nail will bleed if you accidentally cut into the quick. Applying the styptic pencil—or light, direct pressure—will stop the bleeding. Give your kitten time to calm down if you do cut into the quick. Most likely, he will be too upset to finish the grooming procedure.

Combing and Brushing

It will be helpful to have two types of combs—one fine-toothed and one with wide teeth. Comb your

Skunky Encounters of the Smelly Kind

In the unlikely event that your kitten gets into a spraying war with a skunk, you won't have any trouble figuring out what happened once you get a whiff of that unmistakable skunk scent. You also won't have as much trouble figuring out who lost the war.

There are many commercial products on the market to remove skunk stains and smells. However, you can start by shampooing your kitten and then pouring either milk or tomato juice directly on his fur and letting it soak for approximately ten minutes before rinsing well.

kitten to remove any tangles. (Long-haired kittens may need to have tangles removed by hand so the brush or comb does not pull the tangled fur out. Only cut a knot out as a last resort, and be careful not to let the scissors touch the kitten's skin, since it pierces easily.) You can use a bristle brush or a soft acrylic brush to smooth out your kitten's coat, making sure to brush the fur in the direction it naturally lies. Brushing the excess fur from your kitten's coat will

Long-haired breeds like Himalayans require weekly or even daily grooming.

not only prevent him from getting hairballs, it also stimulates circulation and helps keep his coat shiny and healthy looking.

Bathing

Generally speaking, felines and water do not mix well, but starting to bathe your kitten early in life can make all the difference between dealing with a calm and relaxed cat and one who is a ferocious biter and scratcher. You should note, however, that bathing a cat should only be done when it is absolutely necessary, such as if he's been sprayed by a skunk or gotten something oily or foul on his fur.

Put a mat on the bottom of the sink to help prevent your kitten from slipping by giving him more secure footing. You should always keep your hands in the water flow in order to detect any change in the water temperature. (A kitten could rub up against the faucet controls and change the water temperature, leading to serious burns.)

Fill the sink with an inch or two of lukewarm water and, while holding your kitten firmly but gently, lightly wet him. Rub in a small amount of shampoo (your veterinarian can recommend one) and work it through your kitten's coat. Avoid getting any water or

Most cats would prefer to be brushed rather than bathed, so bathing should be done only when necessary.

Groomers Who Make Housecalls

With our highly active lifestyles, it's no surprise that there are numerous mobile pet groomers. A fully equipped van will pull up to your front door, and the groomer will give your cat a professional grooming in no time. Ask your local veterinarian or check your phone book for a groomer who makes housecalls in your area.

shampoo in your kitten's eyes. Rinse your kitten with warm water until all traces of the shampoo are gone from his coat.

Dry your kitten with a soft, fluffy towel and keep him away from drafts until he is completely dry.

Ears

Be sure to clean your kitten's ears whenever you groom him. The ears should be clean and free of any dark, waxy residue. Swab the ear area with a clean cotton ball dampened with a touch of baby oil. Felines have very sensitive and delicate ears, so handle them carefully. Never stick anything into the ear canal, because you could damage the eardrum. Consult your veterinarian if you think your kitten has ear mites (sometimes indicated by ear scratching) or if the ears smell foul. Your veterinarian can provide medication to eliminate these problems.

Clean your kitten's ears whenever you groom him; healthy ears should be clean and free of any waxy residue.

Eyes

A kitten's eyes require very little attention during grooming. If he has runny eyes, gently wipe any discharge away with a cotton ball dampened in warm water. However, if your kitten's eyes are constantly runny and have a discharge, you should check with your veterinarian.

Teeth

Brushing your kitten's teeth is an important part of the grooming process. Good dental hygiene is important throughout your cat's life. Plaque and tarter build up on the teeth over time and should be removed. Once again, starting your kitten now will get him accustomed to the procedure.

Before you start to brush your kitten's teeth, gently open his mouth and examine his teeth. The gums should appear pink and healthy. Any signs of swollen or bleeding gums, or broken or loose teeth, should be brought to the attention of your veterinarian immediately. Rub your kitten's teeth gently with a cat toothbrush and cat toothpaste recommended by your veterinarian. Never use human toothpaste—it can make your feline ill.

Professional Groomers

Choosing a groomer for your kitten is much like choosing a hairdresser for yourself. Most people do not just randomly pick a phone number from the yellow pages. You usually start asking around and finding out whom your friends and family take their business to. The same should be true when you look for a groomer for your kitten. Your veterinarian should be able to recommend a good groomer whose work they have witnessed on their own patients. Make sure you check the references on any groomer you decide to use.

Health Care for Your Kitten

Choosing a Veterinarian

Choosing a veterinarian for your kitten is a big responsibility. It is important that you like and trust your kitten's veterinarian so that the three of you can form a long and honest relationship. Your veterinarian will play an important role throughout your cat's life, so take the time to search for the right doctor.

One of the first things you should do for your new kitten is to bring him to a veterinarian for a first exam to make sure that he's healthy. Your kitten will also need all the necessary vaccinations and a complete physical once a year. Your veterinarian should

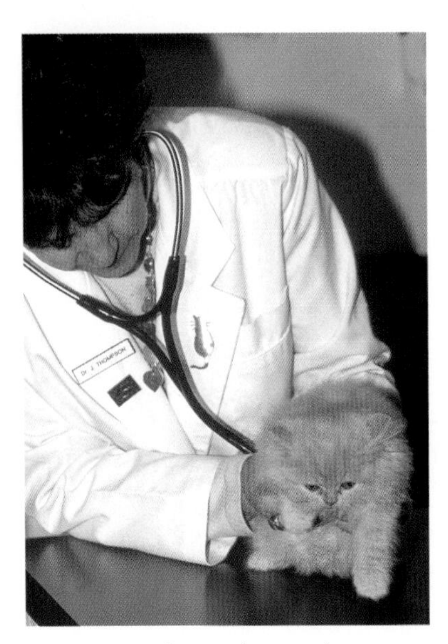

You may prefer to take your kitten to a feline-only veterinarian, who will have specialized equipment and knowledge to better keep your kitten in good health.

check for parasite infections, run blood tests, urine/feces cultures, and any other tests or special examinations that may be needed.

One of the best ways to find a good, reputable veterinarian is to ask other cat owners, friends, and family members whom they take their cat to. When you have found a veterinarian you are thinking of using, schedule a visit and arrange for a tour of the facilities. The veterinarian's office and examination room should be clean and neat. Be sure to observe the staff and see how they interact with the patients and their owners. Everyone working in the veterinarian's office should treat all of the animals

Purring Can Mean Pleasure or Pain

Just because your kitten is purring doesn't mean he's happy, as most people believe. While it is true that felines will purr contentedly while being petted by their favorite human, they will also purr in other situations. A mother cat will purr while her kittens are nursing, and cats that are frightened or sick sometimes purr as well. Scientists have yet to determine why and how cats purr, although they believe the sound is produced by a second set of vocal chords in the back of the cat's throat. If your kitten is purring without an apparent reason for happiness or contentment, you might want to check with your veterinarian.

The Cat's Jumping Ability

Have you ever wondered why, when you put a feline in a big box or put up a tall fence that you just know he can't hurdle, your cat always manages to surprise you? Cats are magnificent jumpers, and most are capable of jumping five times their body length. That's the equivalent of a six-foot-tall human jumping an astonishing 30 feet in the air!

gently and with respect. And, of course, take some time to talk to the veterinarian.

Feline Specialists

An alternative to taking your kitten to a standard veterinary practice is to take him to a cats-only clinic. More and more of these facilities are opening up around the country. Veterinarians who specialize in felines are often able to have more specialized equipment than that found in facilities that treat a variety of animals. Another bonus to your kitten is that the reception area of a cats-only practice is more relaxing—no barking dogs!

With any veterinarian you choose, make sure you check the doctor's credentials.

Neutering or Spaying

Once of the first topics you should discuss with your veterinarian is spaying or neutering. At one time, it was the norm to alter any kitten at approximately six months of age, but as technology has progressed, more and more experts are agreeing that these simple surgeries are suitable for younger kittens. Your veterinarian should advise you of your options and discuss them with you.

The process of neutering (castrating) a male kitten is a simpler procedure that spaying a female and therefore is less expensive. If the operation is performed early enough in a male kitten's life, he

Stop the Spraying Before It Starts

Have your male cat neutered as early as your veterinarian deems possible. Once your cat has started this marking habit, neutering may not stop him from continuing this behavior for the rest of his life.

will most likely never start spraying (in which the male backs up and sprays a pungent-smelling urine on an object). This territorial signal is a sign to other male cats that this area is already occupied and also serves as a warning that a fight will ensue to capture the attentions of any female in heat in the territory. Since neutering a male after he has started to spray does not guarantee that he'll stop (felines are creatures of habit), it is important that you neuter your male kitten before spraying becomes a problem.

As an educated feline owner, you must never fall victim to the thought that you should let your female experience the joys of motherhood at least once in her life. Although a female cat with young kittens usually makes an excellent mother through instinct alone, her attachment to her kittens disappears after they are weaned. Although it may seem unfeeling and cold, a mother cat wants nothing to do with her offspring the moment they are ready to fend for themselves. Maternal yearnings and longings are not part of a feline's overall plan.

Neutered cats are generally happier, healthier companions.

It is much better, both physically and emotionally, to spay your female as early as your veterinarian allows—usually between six to eight months.

Accustom your kitten to a carrier early on, and make sure he doesn't associate it with only unpleasant experiences, such as trips to the vet.

Altering a female kitten is a bit more complicated (and slightly more expensive) than altering a male. Spaying the female involves a small incision in her abdomen for the removal of her reproductive organs. She will be in surgery for less than thirty minutes. Some veterinarians require an overnight stay at the animal hospital following spaying, while others will let the kitten come home—but only under your watchful eye, with the promise to keep activity levels to a minimum until the stitches are removed.

How Long do Cats Really Sleep?

Cats generally sleep anywhere from 12 to 18 hours a day. About three and a half hours of this sleep is called REM (Rapid Eye Movement), which is a type of very deep sleep. The brain is most active during this stage of sleep. The remaining hours of a cat's sleep cycle are spent in a sort of catnapping state—meaning the cat remains partially alert to his surroundings.

Signs of Illness

Common signs that indicate your cat could be ill include:

- unkempt or dirty fur
- runny nose and discharge from the eyes
- difficulty urinating
- suddenly not using the litter box
- haw (third eyelid) exposed
- vomiting or diarrhea
- not eating or drinking for more than two days
- excessive thirst
- wheezing or trouble breathing
- falling over/loss of balance
- cries out in pain when picked up

If you notice any of these symptoms in your cat, contact your veterinarian immediately.

Spaying will not alter your female kitten's disposition, except to make her less nervous, less noisy, more relaxed, and more playful and affectionate.

Vaccinations

Most experts agree that there is no longer a one size fits all approach to vaccinations and that each veterinarian needs to assess each patient for his individual risks. Overall, vaccines do far more good than harm, but not all patients need all vaccines. Currently, most veterinarians are still giving the three-way combination (Feline Rhinotracheitis, Feline Calici, and Feline Panleukopenia [FRCP]) yearly, although some are giving it only every three years.

The American Association of Feline Practitioners (AAFP) and the Academy of Feline Medicine (AFM) list two types of vaccines—core and non-core. The core vaccines are those that the AAFP/AFM recommends for all cats. These include the rabies and the FRCP vaccines.

Risk of exposure is probably the biggest deciding factor when discussing your kitten's vaccination needs with your veterinarian. If your kitten has no chance of being exposed to a certain illness or disease, then vaccinating against the disease is probably not necessary.

The non-core vaccines fall into this category. Your kitten's veterinarian should discuss all the possible scenarios regarding these vaccines with you.

Perhaps one of the most common vaccinations that some veterinarians suggest their patients skip is the Feline Infectious Peritonitis vaccine, except in very exceptional circumstances (e.g., a breeder experiencing an outbreak). The Bordetella vaccine is best reserved for felines who are boarded frequently or who are show or breeding cats.

If you allow your kitten outdoors, it may be best to supervise him so that he doesn't expose himself to unnecessary risks.

Urinary Tract Problems

Urinary tract blockage is more common in male cats than in female cats. Frequent urination, straining to urinate, bloody urine, or crying out when trying to urinate are the most common symptoms. If your cat is having problems urinating, take him to the vet as soon as possible.

Although the Feline Leukemia Virus (FeLV) vaccine is usually only given to cats who are at a realistic risk of acquiring the disease, many veterinarians give it to all kittens, since their lifestyles can change in the years ahead and since felines younger than one year old are at the greatest risk of FeLV infection. In subsequent years, your veterinarian may re-evaluate the risks and decide whether to continue giving the vaccine.

Common Diseases and Ailments

Although there are many illness and diseases that can affect felines, the following are some of those that are the more dangerous and that can be spread from cat to cat.

Feline Immunodeficiency Virus, sometimes called Feline AIDS, has no connection to the human version of the AIDS virus. It only mimics some of the symptoms; hence the name. FIV-positive felines can live with the disease for years after being diagnosed, as long they receive regular veterinary care. Since affected felines have a lowered immune system, they are very susceptible to catching other illnesses and diseases. Although there is no vaccine or treatment for FIV currently available, an FIV-positive kitten can live a happy, fulfilled life as long as he's kept on a regular preventive medicine schedule.

Feline Infectious Peritonitis is quite deadly. It occurs more commonly in felines frequently exposed to large groups of other felines than in those living in single-cat homes. The vaccine for this disease is under scrutiny because some experts are not sure if it offers enough protection against the virus.

The Feline Leukemia Virus (FeLV) is an illness passed between cats by direct contact. The first thing you should do whenever you adopt a new feline is to have him tested for and then vaccinated against this deadly disease. FeLV-positive felines can suffer from a wide range of symptoms, including cancerous tumors, anemia, and a weakened immune system. Keeping your feline up-to-date with his vaccinations can help prevent the spread of this vicious virus.

The Feline Panleukopenia Virus, more commonly known as Feline Distemper, can live in the ground for several years. Your kitten must be vaccinated against this disease. Many felines, especially kittens, who are infected with this virus do not survive.

Upper respiratory infections are especially hard on a kitten. The infection can be airborne (from coughing and sneezing on another feline) and can also be spread by the feline's saliva or runny discharge from the eyes or nose. The virus can also live on a human

Cats spend 30 to 50 percent of their waking hours grooming themselves, so if your kitten seems dirty or unkempt, it may be a sign that something's wrong.

and be spread by petting an infected cat and then petting an uninfected one. As with other serious disease, your kitten should be vaccinated against upper respiratory infections.

Parasites

Fleas and ticks are the most common parasites to affect felines. Be sure to examine your kitten for these parasites every time you groom him, or at least once a week. Indoor cats who have no contact with the outside world run virtually no risk of contracting fleas or ticks. If your kitten does develop a problem, your veterinarian can recommend the best treatment.

Some kittens have worms during the first few months of their lives. This is especially true in the case of strays. Be sure to have your veterinarian check your kitten for worms by testing a stool sample. Most intestinal parasites, including tapeworms, roundworms, hookworms, whipworms, and flukes, can be treated with great success by your veterinarian.

Time is Love

All the preparation, training, and care outlined in this book might seem like a lot of work. Many people think owning a kitten simply means coming home with a little bundle of fluff, placing him in the house with a litter box, and feeding him once or twice a day. There is much, much more to it than that. Take the time to truly know and love your kitten. Play with him, spend time with him, talk to him, and enjoy the process of watching him grow up into a beautiful and loving companion. Love your kitten, and he will love you right back.

Resources

ORGANIZATIONS

The Cat Fanciers' Association (CFA)
P.O. Box 1005
Manasquan, NJ 08736-0805
Phone: (732) 528-7391
Fax: (732) 528-7391
E-mail: cfa@cfainc.org
www.cfainc.org

The International Cat Association (TICA)
P.O. Box 2684
Harlingen, TX 78551
Phone: (956) 438-8046
Fax: (956) 438-8047
E-mail: ticaeo@xanadu2.net
www.tica.org

Fédération International Féline
c/o General Secretary
Little Dene, Lenham Heath
Maidstone, Kent ME17 2BS ENGLAND
Phone: 01622 850913
E-mail: general-secretary@fifeweb.org
www.fifeweb.org

The Governing Council of the Cat Fancy (GCCF)
4-6, Penel Orlieu
Bridgwater, Somerset TA6 3PG
ENGLAND
Phone: 01278 427575
www.gccfcats.com

PUBLICATIONS

Cat Fancy
Subscription Department
P.O. Box 53264
Boulder, CO 80322-3264
Phone: (800) 365-4421
E-mail: fancy@neodata.com
www.catfancy.com

Whole Cat Journal
P.O. Box 1337
Radford, VA 24143
Phone: (540) 763-2925
www.wholecatjournal.com

Cat World
Ancient Lights, 19 River Road
Arundel, West Sussex BN18 9EY
ENGLAND
Phone: 01903 884988
www.catworld.co.uk

INTERNET RESOURCES

The Daily Cat
www.thedailycat.com

Cat Fanciers Website
www.fanciers.com

Show Cats Online
www.showcatsonline.com

VETERINARY RESOURCES

American Association of Feline Practitioners
618 Church Street, Suite 220
Nashville, Tennessee 37219
Phone: (615) 259-7788
E-mail: aafp@walkermgt.com
www.aafponline.org

EMERGENCY RESOURCES AND RESCUE INFORMATION

Animal Poison Hotline
Phone: (888) 232-8870

ASPCA Animal Poison Control Center
Phone: (888) 232-4435
www.aspca.org

Cats Protection
National Cat Centre
Chelwood Gate
Haywards Heath
Sussex RH17 7TT ENGLAND
Phone: 08702 099099
E-mail: cpl@cats.org.uk
www.cats.org.uk

Index